The Universe of Dynamics

**Fearfully and Wonderfully Made
In Awesome Harmony!**

Thomas Yi, PhD
Foreword by Dr. Donald B. DeYoung

Copyright © 2022 by Thomas Yi, PhD
All rights reserved
ISBN:

Dedication

This book is dedicated to my wife, Christine; my first son, Jay, and his wife, Jessica; my second son, John, and his wife, Hana; and my two grandsons, Josiah and Jeremiah.

I pray that this book will not only promote the foundation of God's Word for my family but will also help people who are interested in the beginning of the universe and the Creator's will just and love.

Acknowledgements

It has taken many hours of editing by a number of special people to bring this book to completion.

I would like to thank Dr. Donald B. DeYoung (Former President of Creation Research Society), my creation colleague, Mr. Jaeman Lee (President of Association for Creation Truth), and my engineering colleague, James Kim.

Also, my sincere appreciation goes to my wife, Christine; my two sons – Jay and John, and two daughters-in-law, Jessica and Hana, for many times they reviewed and corrected this book.

Table of Contents

About the Author ix

Foreword by Dr. Don B. DeYoung xiii

Prologue xv

Chapter 1: The World of Dynamics – Inevitable Consequence 1

Chapter 2: Unstoppable Satellites in Space 15

Chapter 3: Our Earth Rushes like a Bullet 27

Chapter 4: The Sun and Galaxy also Rush 35

Chapter 5: Creation of Time-Space-Matter 47

Chapter 6: Formless Matter? 59

Chapter 7: Let there be light! 65

Chapter 8: Light before the Sun 73

Chapter 9: Laws of Creation 81

Chapter 10: What is the End of Matter? 87

Chapter 11: Spinning Earth from the Beginning 97

Chapter 12: Just and Love 105

References 115

About the Author

Thomas T. Yi, PhD

Dr. Thomas T. Yi earned a BE degree in Mechanical Engineering from the Hong-ik University in Seoul, Korea. For his military service, he joined a national defense R & D lab as a research engineer. He developed several precision weapons, including airborne rockets and ground-to-ground missile systems.

In 1988, he resigned from his senior researcher position and came to the United States to pursue graduate study. He obtained an MS degree in Aerospace Engineering and a PhD in Multibody Dynamics (in Mechanical Engineering) from the University of Arizona.

During his studies, he formulated a method for deriving a set of system equations for flexible multibody systems and its computer simulations. Additionally, he developed a generic program for the computer simulations for flexible-rigid mechanical systems. Since then, he has worked first as a research associate and then as faculty at the University of Arizona.

THE UNIVERSE OF DYNAMICS

In late 1996, he took a *position as a* technical specialist at the Case Corporation and was eventually promoted to the roles of senior principal engineer and then engineering manager at CNH (Case and New Holland) Industrial, previously known as FIAT Industrial.

Over the last thirty years, he has published a dozen technical journal papers and conducted hundreds of research projects. Most projects involved driving and solving system equations of motion. Through his technical knowledge and skills, he successfully completed various product developments and innovated quality mechanisms such as suspensions for heavy duty vehicles.

In 2020, he retired from CNH Industrial and took a faculty position at ECC (Elgin Community College) and MCC (McHenry County College). Currently, he is teaching General and Engineering Physics, and Analytic Mechanics such as Statics, Dynamics, and Mechanics of Materials at both colleges.

Apart from Dr. Yi's education and work experience, he is dedicated to his faith in Christianity. In 1999, he and his colleague, Mr. Jaeman Lee, organized the Association for Creation Truth (ACT). As part of his service in this organization, he has been a speaker and writer of many creation focused articles.

In 2001, Dr. Yi completed Creation College and then again in the following years at the Answers in Genesis Creation Ministry. And he joined the Creation Research Society (CRS) in 2005. For over the last 20 years, he has visited many different countries and shared the creation truth to thousands of people.

His fields of interest and research include not only the dynamics of space vehicles, such as satellites, spacecraft, and Mars rovers, but also the planetary dynamics of the solar system.

Foreword by Dr. Don B. DeYoung

I am happy to endorse this writing project by my friend Thomas Yi. Dr. Yi has expertise in the mathematics of dynamical systems, that is, objects in motion, especially applied to space technology.

In this concise book Thomas shares his engineering interests in a clear manner. Beyond this content, however, the author also points out clear physical evidence for the Creator of the universe. The reader will find examples of intelligent design and a defense of the recent, supernatural Biblical creation of the earth and cosmos. One will also find Thomas' personal testimony of faith, so refreshing and important in our day.

It takes unusual effort to produce the book you are holding. Thanks to Dr. Yi for sharing his heart with readers.

Don DeYoung, Ph.D.
Board Member (Former President)
Creation Research Society

Emeritus Science Faculty
Grace College
Winona Lake, IN

Prologue

As we know, there are so many theories in science. Some theories are very reliable. Because they give us very close realistic solutions. Most theories in engineering are this way, allowing us to apply them into real life. However, all theories around origin issues are not.

We can say that science deals with two things – scientific laws and theories. What's the difference between them? Most scientific laws are fully proven by experiments, but all theories are based on assumptions. Here, notice that all the laws in nature are not man-made, but discovered by scientists. Many laws in science have been found and being utilized.

From here, we can ask these fundamental questions. Who created such natural laws? Who embedded the laws in nature? This book attempts to promote biblical creation from the very beginning. The dynamic world cannot exist without intelligent input. The entire universe reveals God's creation, and all dynamic worlds proclaim this truth.

Most books about the astronomical universe take an evolutionary and secular perspective. They often raise more questions than answer questions because they bring unconfirmed or unproven theories. From a dynamic perspective, this book will explain all things in the universe, excluding all theories.

The first chapter introduces the universe as a dynamic world in which all things are moving, and everything has an assigned movement. It also introduces the law of motion in an easy-to-understand manner. It then explains in detail why the law of motion is an inevitable result.

In Chapter 2, we will come to understand the dynamics of the universe through man-made satellites. The motion of each satellite can be explained without exception by the laws of motion, also called the law of cause and effect. Therefore, it makes clear that the dynamic world is the result of wisdom.

Moving forward, Chapters 3 and 4 deal with the motion of the macro world—the Earth and its moon, the solar system and the galaxies in the same sense as the man-made satellites.

Chapters 5 through 7 introduce an order of creation consecutively according to the first three verses of Genesis chapter one. Through these chapters, we will see that the Scripture makes sense and bears no conflict with scientific facts.

Chapters 8 through 11 continue with evidence of how the Bible and experimental research are free of conflict. In particular, Chapter 10 deals with a fundamental question – What is the end of matter? Author briefly summarizes the outcomes of research on elementary particle physics and experimental quantum physics.

The final chapter covers some fundamental questions: who is the Creator? The purpose of creation? The will of the Creator?

Each chapter shares that the motions of all objects are inevitable and have intentional results. The results of advanced experiments and scientific discoveries do not conflict with the Bible, but rather confirm various Bible verses about creation and the universe of the Bible as facts. New scientific findings occur every day, but all discoveries related to Scripture will be confirmed to be true.

It is a great honor for me that the former president of the Creation Research Society (CRS), Dr. DeYoung, a man with real talent for writing many creation books, has written the foreword for this book. Dr. DeYoung is the one who the Lord used through his humble research and publications to bring me into the creation ministry.

References in the back are included for documentation of thoughts and for further study. May this book help you to understand the details of the wonderful and fearful dynamic world as they declare the glory of the Creator God!

Chapter 1:
The World of Dynamics – Inevitable Consequence

The universe can be compared to a big house where all things dwell. However, no one knows its size and it is far beyond human imagination. We often ask whether this unimaginable universe was created by itself or out of someone's intention. As far as we know, this universe is strictly coordinated, finely tuned and controlled. These can be easily confirmed directly through scientific findings and facts.

Everything Moves!

Throughout the known universe, everything moves. Anything that appears to be firmly fixed can be measured. There is a difference in the degree of movement, but they all move without exception.

Everything seems to perform in a natural movement with different frequencies and magnitudes, respectively. We can say that this universe is filled with dynamic energy, that is, kinetic energy. The question then is: Why is everything moving?

Let's consider some examples. The Earth rotates by itself 360 degrees every day while orbiting around the sun. However, the earth's only satellite, the moon, always shows us the same face and revolves around the earth every month. Simultaneously, the earth orbits the sun with the moon.

The speed of both the earth and the moon when orbiting around the sun is about 30 km per second, which is much faster than the speed of any bullet. They run a fixed course around the sun and return to their original point in 365 days.

Our solar system consists of eight planets and most of the planets have their own satellites. Each planet with its respective moons has a unique orbit. Their movements in space cannot be stopped even for a single moment. They all do breathtakingly massive scale movements without a break.

Scientists say that the Milky Way galaxy, to which our solar system belongs, is also orbiting outer space. Our galaxy has a complex gravitational motion within its own "local group" of 30-50 galaxies. This will be covered in more detail in later chapters. It is also believed that the countless number of stars within each galaxy are moving much faster than the speed of any bullet. All of these are moving without rest. It is a truly dynamic world!

What about the inner world of matter? It's not much different! Through advanced high-energy experiments, many different types of subatomic particles have been discovered and their weird properties have been revealed. Research simply shows that microscopy is also a dynamic world.

In the well-known standard particle model, electrons orbit a central nucleus, and the electrons transfer energy at the speed of light. Here, we notice that light is also an electromagnetic wave energy traveling through space. Similarly, the inner world of matter is full of dynamic energy. It is also a dynamic world! In the next few chapters, we'll cover this in more detail to get a feel for how dynamic this world is.

Why is Everything Moving?

We can say that the study of dynamics or mechanics was initiated by Galileo Galilei (1564-1642). At that time, he introduced the concept of velocity and acceleration (rate of speed change) mathematically and confirmed it by experimenting with falling objects. He accurately understood the orbital motion of the planets of the solar system.

There are some reports that he faced great resistance from the authorities of the church, who believed in the theory that the earth was fixed, and the sky was spinning around the earth. Often this story is recounted as Galileo being persecuted by the church because of its hostility towards science, but this is a misunderstanding. What he challenged was the wrong position on science, not the church. His efforts at that time were to break prejudice and freely explore the order of the dynamic world.

Laws of Dynamics.

The greatest contributor to dynamics is Sir Isaac Newton (1642-1727). When he was 45 years old, he published a book known as "Principia." In this book, he introduced three laws of motion.

Newton's first law of motion states that all objects resist changes in motion. Everything in the universe rejects change. In other words, if an object is stationary or moving, it tries to maintain that state.

As we experience on a shuttle bus, passengers fall backwards when the bus quickly starts from rest, and fall forward when it stops abruptly. It is not possible for a driver to stop a moving car immediately as shown in the figure below. This is because everything tries to maintain its current state and resists change. This is the so-called law of inertia.

M. Hassan, stockvault.net, CC0 1.0

The second is the law of acceleration. Here the acceleration is the rate of speed change. When force is applied to an object, the object undergoes a change in velocity. It's natural! Here, the change in velocity in time is acceleration. Newton's second law of motion explains how the size of the force is related to the acceleration of the object.

THE UNIVERSE OF DYNAMICS

In this case, the force applied to the object is equal to the product of the mass (m) and acceleration (a) of the object, or Force = Mass x Acceleration. According to this law, we can establish an equation of motion. By solving the equation, it is possible to know exactly how the object moves as a result.

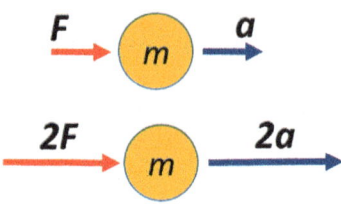

F = ma
The applied force equals to mass times acceleration.

Newton's third law states, "For every action, there is an equal and opposite reaction." When two objects interact, they apply forces to each other of equal magnitude and opposite direction. When a force is applied, the two objects exert a force of the same magnitude in opposite directions.

For example, in a baseball game, the moment a bat strikes a ball, both the bat and ball undergo the exact same impact. It's because of this third law that a rocket launched into space is propelled as shown in the figure. For every action in the universe, there is an equal and opposite reaction.

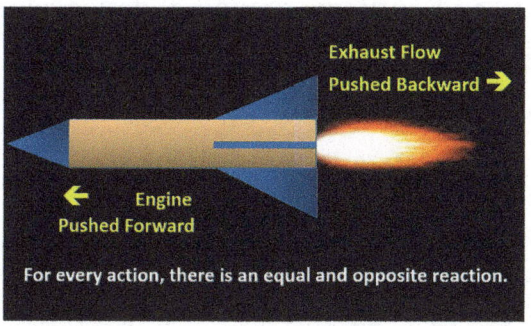

Everything Obeys the Laws!

Newton's laws of motion are indispensable in designing machinery such as satellites, spacecraft, airplanes, automobiles, etc. These laws apply precisely to not only people's daily activities but also macroscopic planets and their satellites without exception.

I majored in multibody dynamics, and for over 30 years have applied Newton's laws of motion in performing various research projects. When designing airplanes, satellites, space vehicles or ground vehicles, as well as any moving machines, equations are established using these laws of motion. The equations are solved with a high-speed computer, and we can visualize the numerical results. We can also see the dynamic behavior of the product through animations. Then, we further analyze the results, and optimize the initial design to improve the performance. These days, it is impossible to think of a better design without the application of these laws of motion.

The introduction of the laws of motion and its application in this chapter is not intended to deal with complex equations. Its intention is to show that the movement of the physical world can be clearly expressed and explained by these laws of motion. No hypothesis or additional theory is required to apply this principle. The laws of motion are the law of cause and effect revealed as a consequence. How can we send a spacecraft or a rover to Mars if we're not sure of these laws of motion?

Who turned the spinning top?

There is no coincidence as far as dynamics are concerned because we can't imagine any movement without a cause. For example, look at a spinning top, who can claim that it started spinning by itself ? It is logically practical to say that this dynamic world is an outcome of an intended input.

Law of Motion Law of Causality!

Isaac Newton seems to have possessed deep insight into the dynamic creation of the world. He thought of the universe as a mechanism like a gigantic clock interconnected with complex gears.

In understanding the universe, there was no need to consider coincidence at all. In his view, everything about the universe would inevitably be understood. He was convinced that all the mysterious phenomena of the universe could be explained by interacting objects and forces.

Isaac Newton made various references about the creation story of the Bible in the second edition of *Newton's Principia: The Mathematical Principles of Natural Philosophy*, a confession on the solar system:

> *This most beautiful system of the sun, planets, and comets, could only proceed Pom the counsel and dominion of an intelligent and powerful being... This Being governs all things, not as the soul of the world, but as Lord over all; and on account of his dominion he is wont to be called Lord God "pantokrator," or Universal Ruler...* [1]

It is clear that Newton was a man who was convinced of God's creation and like many other pioneers, was a scientist who clearly demonstrated that true science begins with a humble recognition of the Creator.

Holistic Science

Today's science (education) does not take a holistic approach. We teach the laws of science and their content, but we are silent about the reasons and meanings of why such laws exist and who created them.

The goal of science should be to find facts in nature. However, today's science is totally limited by naturalism and interprets all the scientific data while excluding the transcendent. We can say that the definition of science has been changed. It is biased to accept scientific knowledge that only fits the naturalistic worldview.

How interesting would it be if we were to acknowledge the Creator in science class and find the hidden secrets, that is, the wisdom of God in the laboratory? All experiments would be more serious, more meaningful, and more humble!

An Inevitable Conclusion

The universe is a dynamic world. This is an inevitable consequence so that it can clearly be explained by the law of causality. Of course, naturalists can tell other stories like the big bang theory or another theory. Briefly note that the big bang theory is an explosion and a rapid expansion of an initial "kernel" of mass energy.

However, all explosions only increase disorder, and there is no natural process of increasing order. It is far from a strict dynamic system. It is convincing and reasonable to consider the Creator as the provider of its cause in the dynamic system of the universe.

All scientific laws governing creation are perhaps evidence of the Creator's deep trust in his creation. If we look closely, we will be convinced, and in the end, we will be able to feel his workmanship (Rm 1:20).

Dynamics are essential to every aspect of our lives, including our body's blood circulation, brain functions, breathing, and muscle activity and so on. This is the result of the Creator's wisdom in knowing all things and maintaining all the necessities that exist on the cosmic scale and small micro-worlds. The dynamic world of creation truly reveals the glory of the Creator and the care he has taken in his creation. In the next chapter, let's examine how dynamic this world is through spacecraft and satellites.

"For since the creation of the world God's invisible qualities—his eternal power and divine nature—have been clearly seen, being understood from what has been made, so that no one can excuse." (Rm 1:20)

Chapter 2:
Unstoppable Satellites in Space

In the previous chapter, we began an exploration of how dynamic this world is. This universe is truly a dynamic world. We will now take a look at this dynamic world through satellites and spacecraft.

Man-made Satellites

With the successful development of large rocket engines in the last half century, space exploration has become a significant symbol of national technology. Even today, the competition between countries remains very hot. Each country has launched numerous satellites for various missions such as for communication, weather, espionage, and more.

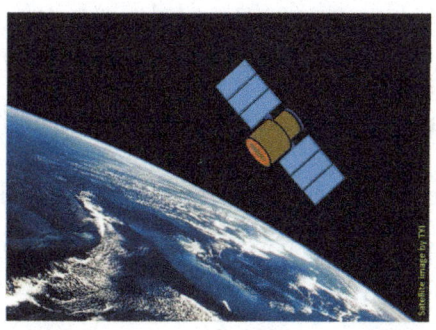

At this moment, there are satellites flying in orbit around the earth without stopping, performing their assigned missions. Just like an airplane flying in the sky, it cannot be stopped in the air. Satellites do not have the ability to stop in space even for a second. They all fly exactly according to the laws of motion. There are no exceptions at all!

None of them have been done by themselves or by chance. Needless to say, when a machine is no longer operational, the Earth's gravity causes it to fall into the atmosphere, burn up, and eventually fall to the ground. There are no exceptions. All things in space are given movement, and they all obey the laws of motion.

Finely Tuned Satellites

When a spacecraft is launched from the ground, it rises to a certain height to enter the target orbit and then maintains that orbit. At that moment, the spacecraft is controlled at a certain speed so that the gravitational force (between the earth and the spacecraft) and the centripetal force due to the acceleration of the spacecraft are equal. Then, the spacecraft maintains this same speed during flight.

A critical speed is defined when the spacecraft's centripetal and gravitational forces are in a tight balance and run in the circumferential direction.

If the spacecraft is slower than the critical speed, it will be in continual free-fall towards the earth, and if it gets faster, it will move away towards outer space. So, engineers must accurately calculate the required speed for every spacecraft and set it to this speed.

As long as the speed is maintained and if there is no atmospheric resistance, the spacecraft will continue to orbit the same trajectory forever just like the earth endlessly orbits around the sun.

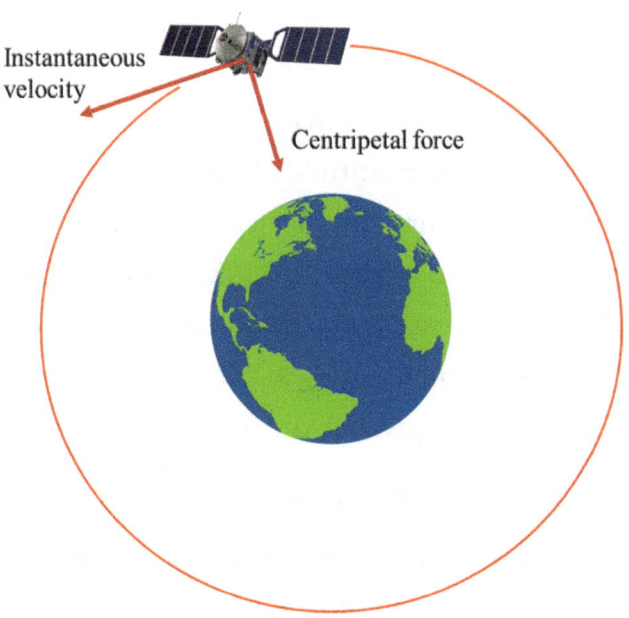

Balance of centripetal and gravitational forces

Manned Satellite ISS

Among satellites, the ISS (International Space Station) is a people-on-board spaceship. The ISS orbits our Earth's trajectory at an altitude of 408 km. Once the spacecraft reached its target orbit, the first thing it had to do was to find the specific speed to maintain the altitude. This is the specific speed at which gravitational and centripetal forces are exactly balanced.

For the ISS, it is calculated at about 8 km per second and is controlled at this speed and maintains the same orbit. As a result, the space station orbits the earth every 90 minutes.

One thing to note here is that there is a certain amount of atmosphere at the altitude of the ISS. The space station is subject to air resistance and friction, which can lead to a change in altitude. It is often necessary to correct the trajectory using the propulsion engine. If there is no air present, then the space station will orbit forever given its trajectory just like the moon.

There are some satellites that maintain a fixed position above the earth. These are commonly known as geostationary satellites or geosynchronous satellites. They appear to be in a fixed position when viewed on Earth because they are located on the equatorial belt. These satellites are tuned to have the same circumferential velocity as the earth's rotational speed at 36,000 km above the earth. They also orbit the earth at a specific speed of 3 km per second in a 24-hour cycle.

Like all planets of the solar system, all satellites are orbiting at a given speed in space. So far, each country has developed and launched various kinds of satellites, and thousands of satellites are currently orbiting the earth now. Everything is engaging in a dynamic movement.

A Day on a SpacecraL?

The ISS is the only manned spacecraft orbiting planet Earth. The fact that there are people aboard the ISS spacecraft is very intriguing. This spacecraft runs faster than a bullet and travels around the earth sixteen times a day. Nevertheless, the crew do not feel the flying speed and spend each day in the spacecraft.

In the spacecraft, they experience zero gravity due to the balance of the earth's gravity and centripetal force. No matter how big or heavy an object is, they cannot feel its weight like on Earth.

Weightlessness Zero Gravity?

Weightlessness is complete or near-complete absence of the sensation of weight. This is also called zero-Gravity or "zero G-force."

The phenomenon of "weightlessness" occurs when there is no force of support on our body. When our body is effectively in "free fall", accelerating downward at the acceleration of gravity, then we are not being supported.

For better understanding, let's consider a high speed elevator. If we stand on the scale in an elevator accelerating upward, we feel heavier because the elevator's floor presses harder on our feet, and the scale will show a higher reading than when the elevator is at rest. On the other hand, when the elevator accelerates downward, we feel lighter. When the downward acceleration (**a**) reaches the same magnitude as the gravity (**g**), we will undergo weightlessness.

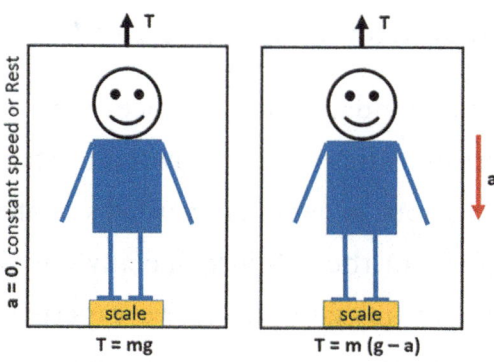

Earth-orbiting astronauts are weightless for the same reasons that riders of a free-falling elevator are weightless. When a satellite is moving with constant speed along the orbit, the satellite as well as the person inside it, both fall towards the earth with gravitational acceleration. Since they are moving with constant speed and continually changing their direction along the orbit, they are undergoing the centripetal acceleration towards the center of the earth. The gravity pulls them down while the acceleration force seems to pull them up. At a certain speed, these opposite forces balance each other out, making them feel a sensation of weightlessness.

THE UNIVERSE OF DYNAMICS

The International Space Station and other satellites are designed to stay in orbit, neither falling to the ground nor shooting off into space. When astronauts are in orbit, they are in free fall, and are weightless.

No Top or Bottom in Space Station!

There are no standards or references for what is the top or bottom. They can sit down and stand in any direction without experiencing any discomfort. Their days are spent in the ISS just like any day on Earth. This is exactly the same thing we experience as the earth orbits the sun at a tremendous speed, but we are not aware of it.

ISS: Zero Gravity

The ISS spacecraft can accommodate up to six people. Their days are spent like any day on Earth. Each person has a busy schedule with assigned missions to complete. They typically wake up at 6 a.m. in the morning, check their daily schedule, eat their daily meals, take a shower, read a book or watch a movie, and then go to bed at 9 p.m. They spend about six months in space and then return to Earth after their mission is completed.

The Length of a Day in Space Station?

One might ask, "How do they set a day on the space station?" One day on Earth is the cycle of the earth's rotation, which is 24 hours on the standard clock. So, a day on Earth is 24 hours. The moon orbits the earth at about an altitude of 380,000 km. The period in which the moon orbits the earth is 29.5 days, which is the definition of a month.

Meanwhile, the space station orbits the earth every ninety minutes. Should this period be considered a day? Of course, not! A day on the space station has nothing to do with the earth's rotation, but they use 24-hour days just like on Earth.

It should be noted that as far as time is concerned: **morning** and **evening**, **day** and **night**, **a day** and

seasons, and **a year**, all origins of these are not defined by humans. Who or what determined these, and how? The answer to all of these questions about our origins can be found in the book of Genesis in the Bible.

"He separated the light from the darkness. God called the light 'day,' and the darkness he called 'night.' And there was evening, and there was morning—the first day." (Gn1:4b-5)

And God said, "Let there be lights in the expanse of the sky to separate the **day** from the **night**, and let them serve as signs to mark **seasons** and **days** and **years**." (Gn1:14)

A Lesson from Space Explorations

People have benefited in numerous ways through space explorations. For example, we have more accurate weather forecasts with the use of meteorological satellites. Wireless communications are possible everywhere through the power of communication satellites. Through GPS satellites, it is possible to track our location wherever we go. How convenient has it become to drive a car with a GPS map or to operate an airplane? It is hard to imagine everyday life without satellites.

Above all, one thing we have learned from space exploration is that the universe is a very dynamic place.

As mentioned earlier, dynamic systems are accurately expressed and explained by the laws of motion. When it comes to dynamics, it is impossible to explain it by itself or by chance. As you know from a spinning top, there was an intentional input from the beginning. Likewise, as every single satellite is the result of human wisdom, the dynamic motion is an inevitable result and consequence of wisdom.

The earth, moon, sun, and all of the planets covered in the following chapters are operated following strict order like a man-made spacecraft. All of them run in different tracks at a given speedand direction. So, who determined the orbits, critical speeds, and directions of all the planets and moons?

From a dynamics point of view, someone has to accurately calculate these variables and set them into each planet and its moons. These facts disprove the big bang theory that the universe accidentally exploded on its own and expanded itself to form the current state.

It is very surprising that the Bible, which describes the creation, records very interesting facts that even cutting-edge scientists in the twenty-first century do not know. In the next chapter, we will see how dynamic and special is the planet earth we live on.

"Praise him, sun and moon, praise Him, all you shining stars. Praise him, you highest heavens and you waters above the skies. Let them praise the name of the Lord, for he commanded and they were created. He set them in place for ever and ever, he gave a decree that will never pass away." (Ps 148:3-6)

Chapter 3:
Our Earth Rushes like a Bullet

Someone may think our home, the earth, is standing still, but it isn't. The earth is spinning like a giant top. Then, why can't we feel the earth turn? The answer is simple. Because we are so small relative to the size of the earth. But we all know it does turn because that's what gives us our day and night.

In the morning when the eastern sky brightens, we know that our position on the earth is turned toward the sun. In the evening, when the sky grows dark, we know that we're turned away from the sun. The time it takes the earth to make a complete 360 degree turn is about twenty hours, which is one full day and night.

Water Planet Earth
(Taken from Apollo 17, 1972)

THE UNIVERSE OF DYNAMICS

Water Panet

We know the earth is a giant ball of water. These days we can see real photos of our planet taken from a spacecraft. These photos confirm that the earth we live on is a huge object that is round and mostly covered with water.

Scientists estimate that the earth is about 6400 km in radius and weighs about 6 x 1024 kg. As we all know, the earth doesn't just spin; it also runs around the sun. Nobody today will deny it!

Spinning and Rushing Earth

Right this very moment, our earth rotates itself fifteen degrees every hour in a 24-hour cycle and at the same time it orbits around the sun at a tremendous speed. The orbiting speed is about 29.8 km per second. This is at least thirty times faster than the speed of any high-speed bullet. Our Earth is like a high-speed bullet, spinning and rushing around the sun for 365 days, 5 hours, 48 minutes and 46 seconds until it returns to its original position. According to the Bible, this is the exact definition of a single year.

Rotation

15 degrees per hour

Revolution

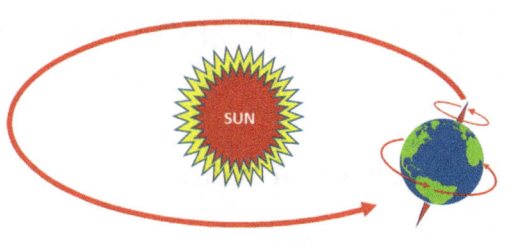

30 km per second

The moon, the only satellite of our Earth, has the same rotational and orbital period. Therefore, we can only see one side of the moon from the earth. That's how it was set up. The cycle of the moon orbiting the earth is 29.5 days, which is equivalent to one month. It is the wonderful wisdom of the Creator that allows the invisible time to be counted through visible things.

And God said, "Let there be lights in the expanse of the sky to separate the **day** from the **night**, and let them serve as signs to mark **seasons** and **days** and **years**,' (Gn1:14)

As explained in the previous chapter, the circumferential speed is determined so that the gravitational and centripetal forces among the sun, the earth and the moon are in tight balance. Again, this speed is very critical for each planet and it obviously cannot be explained by coincidence. Due to the critical speed, they can continue to maintain orbital motion endlessly.

A Planet of Beauty and Wonder

Astronauts who have gone out to space consistently share that there is an inexpressible beauty when looking at the earth from space. They are capable of feeling the magnificence more than anyone else while traveling around the earth.

Guy Gardner, who has traveled on the space shuttle Atlantis, stated that his first words were "beauty of the earth" and the second ones "the wonder" while looking at the earth from space. He confirmed that "Our planet Earth is functionally complete and aesthetically perfect! Such a huge scale and perfect planet cannot be created by itself accidently and spontaneously."[2]

A Treasure Trve of Rare Materials

Scientists understand that the most common element (material) in the universe is hydrogen (73%), followed by helium (25%). They say that 98% of all space matter is hydrogen and helium, while the other 2% is made up of all the rest of the other elements in our universe.

Let's consider our sun. The sun is a gaseous body composed mostly of hydrogen and helium. Scientists estimate that all stars are similar in structure to our hydrogen and helium filled sun. [3-4] Except for the lightest elements, hydrogen and helium, other elements are very rare in the universe.

All the stars in the universe are like the sun and are fundamentally different from our Earth. But our Earth we inhabit is very different and unique, particularly because there are a lot of complex elements on Earth. All the materials of our Earth are very rare and special. We can say that our Earth is a treasure trove of rare materials.

Standing on the Tip of Sharp Blade!

Not only from a material point of view, but our Earth is precisely coordinated with a number of factors. For example, its distance from the sun, rotational speed, orbital speed, tilt angle of the axis of rotation, its size and weight, etc. are all perfectly attuned to be suitable for sustaining life on Earth.

This means that the various laws of physics are tightly interlocked with each other. They are precisely balanced just like standing on the tip of a sharp blade. All scientists agree on these facts [5-6]. If we list the elements necessary to sustain life on Earth, there are so many that these facts cannot have possibly happened on their own or by accident.

Why Oxygen and Nitrogen?

Let's consider the atmosphere on the earth. The atmosphere is composed of 78% nitrogen and 21% oxygen. Here we might have some questions. Why does the earth's atmosphere contain 78% nitrogen, which is an essential element for plant growth, and 21% oxygen, which is essential for breathing animals? Why is the composition of this ratio always maintained? And why is nitrogen and oxygen uniformly mixed together in the atmosphere?

These are very challenging questions for scientists to answer. Nobody knows and there is no answer in any reference book. However, the book of Job in the Old Testament tells us that God the Creator determined the weight of the air and the amount of water as:

"For God, he views the ends of the earth and sees everything under the heavens. When he established the force of the wind (air) and measured out the waters..." (Jb28:24-25)

A Very Special Planet!

Yes! The earth is very special, and we live in a treasure trove. Regardless of whether we recognize it or not, even at this moment, the earth runs as fast as a bullet and performs its given mission. Perhaps this may be a consideration of the wise and almighty Creator? God is watching and holding everything on earth so that we can all live in peace on this special planet.

THE UNIVERSE OF DYNAMICS

In the next chapter, we will take a closer look at how dynamic the solar system and galaxies are.

"Where were you when I laid the earth's foundation? Tell me, if you understand. Who marked off its dimensions? Surely you know! Who stretched a measuring line across it? On what were its footings set, or who laid its cornerstone?" (Jb 38:4-6)

Chapter 4:
Solar System and Galaxy also Rush

The sun is an enormous sphere ball of hot glowing gas. To humans, the earth seems absolutely too big, but it's small relative to the size of the sun. The sun is much bigger than a million earth. The sun is in a planetary system with eight planets, so called the solar system. (Note that in 2006, the International Astronomical Union downgraded the status of Pluto. Pluto is no longer the ninth planet, but rather a "dwarf planet.")

All eight planets are orbiting around the sun and there are about 168 satellites that so far have been discovered orbiting each planet. There are also about 1-2 million asteroids with a diameter of more than 1 km that form a belt between Mars and Jupiter. Meteorites and comets are also members of the solar system.

The planets around the sun are divided into four inner planets and four outer planets. The inner planets consist of: Mercury, Venus, Earth, and Mars. The four outer planets are Jupiter, Saturn, Uranus, and Neptune. The inner planets are all solid masses, and the outer planets are all gaseous planets.

THE UNIVERSE OF DYNAMICS

As shown in the following composite picture, all of these planets are very different in size and are constantly running along their given orbital trajectory at a given speed.

Solar System Composite Picture

Within the solar system, all the planets circle around the sun, and all the moons in turn circle their respective planets. Gravitational attraction causes all the objects to move like precision clockwork.

Interestingly, all of them have a planar motion on a huge disk. All the planets orbit the sun in the same direction. When it comes to the rotation of the planets, they all rotate in the same direction as well except Venus and Uranus. Nobody knows why these two rotate in the opposite direction. It is very unusual and peculiar!

A Solar System in Motion

All of the planets continue to orbit the sun. They can't stop for a moment in space. As shown in the table below, each planet has a given orbit and rotation speed. So, both rotation period and orbital period are determined for each planet. In other words, there is a fixed time for each planet to rotate and return to its original place. Note that the earth's orbital speed varies continually, faster at perihelion and slower at aphelion. The below table lists the average orbital velocity of each planet.

Planet →	Mercury	Venus	Earth	Mars	Jupiter	Saturn	Uranus	Neptune
Rotation Period	59d	-243d	23.9h	24.6h	9.8h	10.2h	-11h	16h
Orbital Period	88d	224.7d	365.3d	687d	11.9y	29.5y	84y	164.8y
Orbital velocity (km/s)	47.9	35	29.8	24.1	13.1	9.6	6.8	5.4

(h=earthly hour, d=earthly day, y=earthly year)

Let's consider Mercury first, the closest planet from the sun. It takes 59 days for it to spin on its axis and 88 days for it to orbit the sun. Here the day means our Earthly day. So, one day on Mercury equals 59 Earth days and one year on Mercury equals 88 Earthly days. Likewise,

Venus's one day is 243 Earth days and one year is equivalent to 225 Earth days. One day on Venus is longer than a year.

The Earth rotates itself 15 degrees every hour to complete its rotation in 24 hours a day. And like a bullet, it is running around the sun's orbit and returning to its original position to accomplish 365 days a year. In this way, not only the inner planets but also the outer planets and their moons of the solar system are performing their own missions, respectively. Especially when it comes to talking about the precision of the motion, its accuracy is much more precise than any man-made machines. Needless to say, in terms of scale, this is just beyond imagination!

In chapter two, it was explained that in order for a spacecraft to maintain a specific orbit, it is essential to determine a constant speed suitable for the trajectory. According to the laws of motion, the motion of all planets is an inevitable result and evidence of the law of causality.

How Giant is the Solar System?

Scientists regard the sun as an average-sized star and a unique source of energy for the solar system. A light beam created by the sun takes about 8 minutes to reach our Earth, and 5 hours and 30 minutes to reach Neptune, the last planet in the solar system. In other words, a light beam crossing the solar system would require more than 10 hours for the trip. It is not easy to imagine the size of the solar system itself.

Please bear in mind that the picture of the solar system shown above is a composite picture and not a scaled one. To get an idea of the actual size of the solar system, let's use the following contrast.

Suppose the sun is the size of a soccer ball (220 mm diameter) in the middle of the Soldier Field football stadium in Chicago. Then, our Earth is located about 24 meters away from the center and it is the size of a mustard seed (2 mm diameter) orbiting around the soccer ball. And Neptune is located about 1000 meters away from the center and about the size of a quarter (0.5 mm) of the same mustard seed and also orbiting the soccer ball in the middle of the stadium.

Scaled distance between the earth and the sun
(Enlarged 10 times of the sun and 100 times of the earth)

THE UNIVERSE OF DYNAMICS

As mentioned earlier, the sun is a medium-size star in the universe. The closest star from our Earth is Proxima Centauri. This star is at a distance of 4.24 light-years from our Earth, or four years and almost three months with the speed of light to arrive there. Since the soccer ball in the center of the stadium in Chicago is the sun, the nearest star is another soccer ball on the island of Hawaii located 6,400 km away from it. It is a distance that cannot be drawn even with a significantly reduced scale.

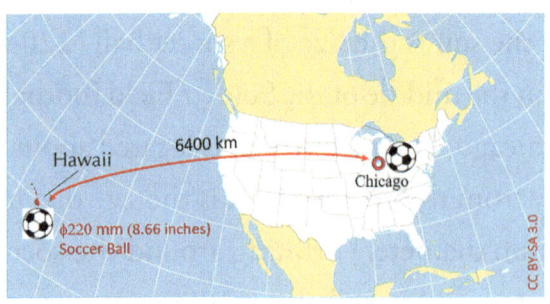

Scaled distance between the sun and the nearest star
(a soccer ball size each)

It is clear to distinguish that the sun and the nearest star are far apart from each other, which is very difficult to imagine. Compared to the distance, their size is nothing more than a tiny speck of dust. However, they are all performing exactly the missions given to them. It is just an amazing and wonderfully dynamic system!

Spinning the Milky Way Galaxy as well!

Let's talk about our galaxy. Our solar system belongs to the Milky Way galaxy. The Milky Way galaxy is thought to have the shape of spiral arms. But the details are uncertain so far.

Today, scientists believe that our Milky Way contains over 200 billion stars, and our solar system is located about two-thirds of the way out toward the outer edge of the Milky Way. The following reference gives an excellent illustration of its size: if the solar system were as small as a coffee cup, then the Milky Way would be the size of North America. [7] It's incredible size is beyond our imagination! Far beyond the Milky Way, other galaxies of stars seem endless.

The study of the Milky Way galaxy evolved following the successful development of large telescopes by British astronomer Herschel (1738-1822). Later, Professor Shapley (1885-1972) at Harvard University integrated study of our galaxy into his courses.

As a result of these studies, scientists have determined the diameter of our galaxy is nearly 100,000 light-years in span, and our solar system is located on an outskirt of the galaxy, about 30,000 light-years from its galactic center. Our planet is not located in the bright central hub of our galaxy, but in a spiral arm.

Spiral Milky Way Galaxy Model

In 1932, with the introduction of radio astronomy, Jansky (1905-1950) and other astronomers proposed a very interesting galactic structure. They said that our galaxy has a helical structure and rotates and runs fast. Subsequently, they estimated 240 km/s of movement speed of our solar system relative to the center of the galaxy. They also insisted that the movement speed of our galactic center with respect to the cosmic background is about 560 km per second. [8]

At that time, all of these findings amazed the academic community. Here keep in mind that we are not to discuss the accuracy of the measurements. But a point to note here is that all scientists claim that the Milky Way, including our solar system, is dynamic. Although there are some differences in measurements, all scientists in recent years fully agree with the fact that galaxies move very quickly. [9]

With these research results, scholars around the world were surprised, but what is even more surprising are the Bible verses. From a dynamic perspective, scientists have come to understand a little about our solar system and our galaxy. However, the Psalms from the Old Testament, written well before 3500 years ago, already recorded the fact that the sun was running, and the sky or heaven was moving.

"In them hath he set a tabernacle for the sun, which is as a bridegroom coming out of his chamber, and rejoiceth as a strong man to run a race. His going forth is from the end of the heaven, and his circuit unto the ends of it: and there is nothing hid from the heat thereof." (Ps 19:6 KJV)

Truly Dynamic World!

In dynamics, all motion is "relative motion." A relative motion absolutely needs a reference. However, no one knows where the reference point in the universe is. The sun apparently moves in a gigantic circuit around the center of the Milky Way galaxy, and the galaxy itself moves with respect to other galaxies.

Dr. Henry Morris explains the reference point in detail in his apologetics commentary book. [10] According to his explanation, "for all surveyors and astronomers, that reference point is the surface of our Earth at the location of the observer. So, David (Psalmist) takes this scientific approach in referring to the sun's motion relative to the Earth, and at the same time, his statement is also correct for any other assumed fixed point, since the sun and the galaxy do actually move throughout the whole universe."

Of course, we do not fully understand the Bible. However, through keywords we can get a glimpse of the movement of the sun and other heavenly bodies.

Scientists say the center of our galaxy has a velocity of 560 km per second. It is flying or shooting through space 560 times faster than the speed of the fastest bullet. This may give us a realistic feel when we think of our Earth traversing the sun's orbit at nearly thirty times the speed of a bullet. Anyway, the earth, the sun, and the galaxy are all traveling through space like bullets.

One question may arise. Although the details of the verse are unknown, how did the psalmist know the movement of our sun and other heavenly bodies? How could he have such knowledge at that time when high-performance observation equipment, radio astronomy technology and computers were not available at all? Incredible!

Wisdom in the Bible

Thanks to the advancements of science and technology, Bible verses that were not understood in the past are confirmed as facts.

Both the Old Testament and the New Testament of the Bible clearly declare that God made the earth especially inhabitable.

"For this is what the LORD says –he who created the heavens, he is God; he who fashioned and made the Earth, he founded it; he did not create it to be empty but formed it to be inhabited—he says: I am the LORD, and there is no other." (Is 45:18)

And

"From one man he made every nation of men that they should inhabit the whole Earth; and he determined the times set for them and the exact places where they should live." (Ac17:26)

Scientific findings that continue to be confirmed support the view that our Earth and solar system in the Milky Way galaxy are very special in terms of their location and their function. [11-14]

Not only through scientific discoveries, but also through the Bible, we can reach a clear conclusion that both the solar system and all galaxies are very dynamic. From a dynamic perspective, this also boils down to the workmanship of the Creator, the Cause Provider, the Subject, and the Super Wise.

The Bible, which records surprising facts, proves that it was written with the inspiration of the Creator. "All Scripture is inspired by God..." (2Tm3:16). From the next chapter, let's take a look at what the Bible says about the universe and its beginning described in the book of Genesis.

Chapter 5:
Creation of Time-Space-Matter

In the book of Genesis, elements that appear on the first day of creation are time, space, matter, and light (Gn1:1-5). Interestingly, scientists also say that the building blocks of the universe are time, space, matter, and light.

In addition, they insist that these elements are interconnected (or combined) together and cannot be split apart. In other words, all of them are one entity or continuum. Also, it is said that matter cannot be imagined without light. We will continue to deal with this issue in the following chapters.

If the universe consists of these four elements, what is the end of the universe? In other words, is the boundary of the universe time, space, matter, and light? This is a question that cannot be answered without transcending this physical world because it deviates from the realm of science dealing with the matter world. To answer this, we can look to the Bible which supports creation beyond time and space.

THE UNIVERSE OF DYNAMICS

"In the beginning God created the heavens and the Earth." (Gen1:1)

It is the first passage of the Bible and the first work God has done. God, who transcends everything, created time as the beginning, the space of heavens and the substance of the Earth at the same time. The reason these three things are expressed together is because they cannot be separated or divided.

The word "creation" used in the first verse of the Bible is "Bara" in Hebrew. This Hebrew word can be used only when something is created out of absolutely nothing. Here, the concept of absolutely nothing (or ex nihilo) means a state without time, space, matter, or light.

In that sense, the end of the universe is a boundary between absolutely nothing and existence. So, trying to discuss the boundary and limit of the universe without knowing the absolute nothingness is nonsense!

Let's look at the description of the Bible furthermore. In the first verse, the creation of time, space, and matter is a state without light. Light first appears in the third verse. Therefore, it is not possible for us to understand the state of time-space-matter without light. This is only possible with an Almighty Creator. This is simply because the space in the first verse is empty, and we cannot imagine a matter whose shape has not yet been determined. We will discuss this formless matter without light in the next chapter.

The 1st Creature is Time.

What is time? We know that time exists in the universe. No one can stop the passage of time. What does it really mean that time doesn't exist? Nobody knows! Because no one has ever been outside of time. Therefore, the one who created time must be outside of the time. And the Creator surely transcends time. That is why the Creator of time is called the Eternal One.

The Apostle Paul professed, "For I am convinced that ... neither the present nor the future . . . nor anything else in all creation, will be able to separate us from the love of God that is in Christ Jesus our Lord."(Rm8:38-39). This means that time is also a creature, and God's love for us is the will of the Almighty who transcends time. The Creator of time is not limited by time. He can see the past, present, and future all at once, all the time.

Since the beginning of history, no one has given a clear answer to the question – "what is time?" In early 5th century, St. Augustine (AD 354-430) analyzed the nature of time and creation. In the 11th book of his Confessions [15], he said that there are three kinds of time in the mind: the present with respect to things that are past, which is the memory; the present with respect to things that are present, which is contemplation; and the present with respect to things that are in the future, which is expectation. He believed that there is neither a past nor a future, only the present. He conceived of the past as the present to be remembered, and the future as the expected present. Therefore, he argued that the past, present, and future do not exist separately.

Here someone may point out about time dilation. That is, time itself can be contracted or stretched experimentally (slightly). Scientists agree that time is influenced by gravity or speed of travel. This idea is fully based on

Einstein's relativity theory. [16] However, such studies about time are far from our daily life.

Of course, we can't see "time" itself. But this invisible time certainly exists and is a component of the universe. Again, time is an independent creature of the universe. It is the awesome wisdom of creation that allows this invisible time to be counted through the visible Earth and Sun. (Gn1:14)

The 2nd Creature is Space.

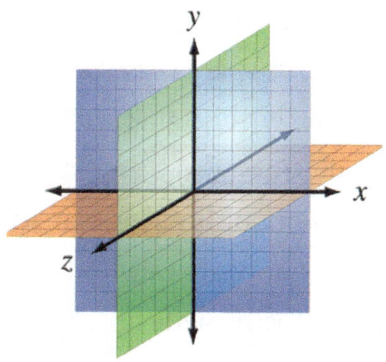

What is space? Space is understood in three dimensions (3-D): front rear, left right, and up down. If this is so, then what is it that there is no space in the universe? This is beyond human imagination. It is because we are already in space, and no one has ever been outside of space. Therefore, the Creator of space must be the one who transcends space. The Bible calls him omnipresent.

Some scientists say that if we remove all the space from our body, there is very little matter left over. Really? This will be easier to understand with an example.

The smallest unit (with properties) of matter is a molecule. Our bodies are made up of molecules, and molecules are made up of smaller atoms. An atom has a central nucleus and electrons circulate around it.

However, both the nucleus and electrons have little mass and are 99.99...% space. Also, are both the nucleus and electrons really substances? Perhaps if we can look into them, they may also be mostly space. In other words, the majority of our bodies are made up of space. This means that not only are our bodies, but all things in the universe are mostly made up of space.

Therefore, since everything is almost space, it is believed that it will pass well without boundaries. Let's consider our bodies and a wall. Since both are made up of space, one can assume that our bodies must easily pass through the wall. But in reality we would get hurt when our bodies hit the wall.

Question? What did our bodies really hit? Most likely, there is no possibility of collision between substances. Strictly speaking, matter does not collide with each other, but forces met in space. The problem is due to forces, not space. The forces acting between molecules, the nucleus, and electrons, and smaller particles in the nucleus are so strong that they cannot pass each other through space. Also, tremendous energy is required to separate them.

Here notice that not only the nucleus and electrons, but all matter has a pulling force with one another. They cannot escape this mutual attraction. In the book of Job, God asked Job the following question.

"who ...when the dust hardens into a mass, And the clods of Earth stick together?" (Jb 38:38 NIV)

"Who will make the clods stick together?" Isn't it really a fundamental question? This is the question that scientists have been contemplating from time immemorial. This verse implies that God himself has already given the character of sticking into space from the beginning. What if God didn't create the attraction that pulls each other? Then, substances cannot exist. God asked this question to Job. Therefore, the one who put this property into space must be omnipotent beyond our imagination.

THE UNIVERSE OF DYNAMICS

The 3rd Creature is Matter.

Now the last question is about matter. Matter has mass, and mass occupies space; therefore, it is not possible to separate matter from space or extract (or remove) space from matter. What is the absence of matter in the universe? There is no way to know this either. Because we are not capable of transcending matter. Therefore, the Creator who says that he created time-space-matter at the same time must be transcendental, all-knowing, omniscient, and almighty.

History is the Flow of Time-Space-Matter

The universe is made up of time-space-matter, and history is the flow of this time-space-matter. This universe is going on a new path that is irreversible, every minute of every hour. We will never experience these three again.

Let's say someone said "we are back home" after a vacation. This is not exactly true because time has already passed, and space has already changed due to the earth's rotation and revolution. Meanwhile, the air has changed, and the house has corroded. Over time, the material has grown different from before. Everything is new. So, rather than returning to the previous home, they are experiencing a brand-new environment.

Therefore, we live a new life that is irreversible every moment and every hour. Some religions believe in the concept of reincarnation, but this is a story that does not fit scientifically because life is irreversible.

The following passage of Scripture points out life exactly as "Just as man is destined to die once, and after that to face judgment" (Heb 9:27). As mentioned in the Scripture, life and history are like an irreversible straight line, not a circle.

In its very first verse, the Bible is asking us about each of our lives: Will you go with the Creator on this irreversible path of life? Or will you go alone?

The very first verse of Genesis tells us about the Creator. Then, is this Creator our Savior? Yes! Who else can be the real savior? The Bible consistently asks us to believe that this Creator alone is our Savior. Because no one can be a true savior other than the Almighty Creator.

Earth, sun, moon, stars, animals, plants, and humans all form a history. There is only one history for all people and all things in the universe. And everybody has his or her own history. Each person can only experience their own. They are all independent, but in the end, they all experienced one history because they did not live in a different time and space. Therefore, what the first verse of Genesis says implies that we all begin our history in timespace-matter.

If someone accepts the first verse of Genesis chapter 1, it is necessary to admit that s/he will accept the next history of creation and subsequent records as it is. If someone accepts the very first verse, but does not believe the other content that follows, then he or she does not know the true meaning of the first verse (Gn1:1).

This may be because they don't really understand what it means to be omnipotent or because they trust man's words more. They might be influenced by the evolutionary idea that from simple to complex creatures have evolved over millions of years.

Irreversible History

History is a flow of time-space-matter and it is irreversible. It is a lie if someone tries to reverse these three things. Therefore, even the Almighty God who created time-space-matter cannot turn this history backwards.

If He tries to turn it back, God himself becomes a liar. When the first man Adam sinned, it was sad, but God did not turn time back. This is because if God turned it back, he would be the one who lied. How does the one without lies turn back to the past? This is not a matter of God's power, but a matter of His character.

Even evolutionists today say that this time-space-matter cannot be separated. This is because they themselves have no choice but to admit scientific laws. However, they insist that this timespace-matter has transformed itself and has become all things in the universe. On the other hand, the Bible records that there is one who transcends the time-space-matter, and that Almighty God created all things according to his purpose.

Chapter 6:

Formless Matter?

"Now the earth was formless and empty, darkness was over the surface of the deep, and the Spirit of God was hovering over the waters." (Gn1:2)

This is the second verse of Genesis chapter one. It is a concrete description of the earth mentioned in the first verse. We can notice that both earths appearing in verses 1 and 2 are the same "Earth" that we live on.

Can Formless Matter Exist?

We can say this passage is a neutral description that the form of matter has not yet been determined. [17] By the way, can we think of any formless matter? No! Impossible! The expression that "the earth was formless and empty" is incomprehensible. It is beyond our imagination. Only the Almighty can say this. Of course, here the earth stands for matter or substance.

THE UNIVERSE OF DYNAMICS

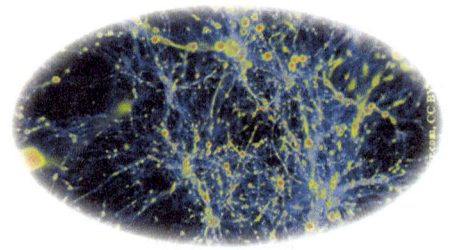

According to the order of the words in the book of Genesis, there was no light yet when God created the formless matter. Light first appears in the next verse 3.

Can we imagine any substance without light? Or can matter exist without light? These days, the correct answer is 'No!' In the quantum world, we cannot think of atomic systems without light. And some scientists in advanced labs have recently proved that "all matter is made of light." [18] There are still many uncertainties, but it makes sense according to the order of creation. We will discuss this in more detail in the next chapters.

In the second verse, it states, "darkness is over the surface of the depth." This indicates that the surroundings are very dark, and the depth depicts a potential or a super energy beyond our imagination.

Here, the depth means a deep water, which is the same word used in Genesis 7:11 at the time of the Genesis Flood known as Noah's Flood. In other words, it means that the earth was created in the presence of water. This is rightly said, "The Spirit of God was hovering over the waters."

Here, we have a glimpse of the image of the Holy Spirit who envelops the earth where his own people will reside and participate in the creation ministry together. Later, his own people will be made in his own image.

The First Substance – Water

Evolutionary scientists imagine that our Earth gradually cooled from a hot fireball over billions of years and turned into a good environment for life. This kind of thinking stems entirely from prejudice because the inside of the earth is hot as the sun. They believe that once upon a time our Earth would have been a fireball.

But the Bible says the exact opposite. The earth was created as a watery planet from the beginning. Here, notice that the word, water, is the first substance mentioned in the Bible. It is the most recognizable and understandable matter for us.

The chemical formula for water is H2O. In other words, a water molecule is made up of two hydrogens and one oxygen. We are not discussing the chemistry of water, but it means that water is not the simplest matter in the world.

There are many simpler things in the matter world: hydrogen, oxygen, nuclei, protons, electrons, quarks, and so on, but God made water right away without steps.

Hydrogen is the most abundant and explosive material in the universe. Oxygen is the third most abundant in the universe and it has the property of oxidizing other substances. Oxygen also has the property of burning all substances as an oxidizing agent; therefore, water is a matter made by burning two hydrogens with one oxygen.

Not from Simple nings, but from Necessities...

As confirmed by high energy experiments in recent decades, the particle world is very unstable when particles (such as nuclei, protons, neutrons, quarks, and leptons) exist independently. We will go over the particle world in detail in Chapter 10.

From a particle physics perspective, the idea that this universe naturally began from such a simple thing is very difficult to imagine. On the other hand, the order of creation according to the Bible is not from simple things, but from necessities.

Evolution scientists want to say that everything started with something simple. This is because, if it started from complexity, it is necessary to acknowledge that creation and its Creator transcend matter. All evolutionary scientists try to unfold the story of all origins, including the origin of the universe, the origin of life, and the origin of animals, etc. This is the nature and limitation of evolution, or naturalism.

However, scientific discoveries and data reveal that simple things cannot be passed on to the more stabilized, more organized, and more complex functions that evolutionists hope for. If some people think that the earth may have started as a hot nebula plasma of gas, then they are influenced by evolutionary geology. There is no geological evidence for this except that the inside of the earth is as hot as the sun.

"Where were you when I laid the Earth's foundation? Tell me, if you understand." (Job 38:4)

The question God asked Job is like a warning against our intention to explain the origin of the universe, excluding God's word and its witness.

Chapter 7:
Let there be light

"And God said, 'Let there be light,' and there was light." (Gn1:3)

This is the third verse of Genesis chapter one. What is light? Physically, light is expressed as having both a wavelike and a particle property. This is an expression of the nature of light, not a definition. No scientist has yet come up with an exact definition of light.

In short, light refers to all areas of electromagnetic "energy" and this universe is full of light energy. Note that light is made up of photons and these photons can possess some of the properties of matter, but light is not matter.

Light includes radio waves, microwaves, infrared rays, visible rays, ultraviolet rays, X-rays, and gamma rays from long-wavelength as shown in the diagram below. Furthermore, it is a comprehensive energy ranging from heat, sound, electricity, magnetism, and molecular interaction as well. [19]

Completion of the Matter World with Light!

Light is the most basic and fundamental form of energy, and it is essential for activating all electromagnetic forms. Therefore, the expression of creating light is most appropriate to say that God who created time-space-matter adds energy. As a result, all formless matter in the second verse of the Genesis one was activated with light energy. In other words, the quantum world was activated with light energy so that finally the universe—the matter world—was completed

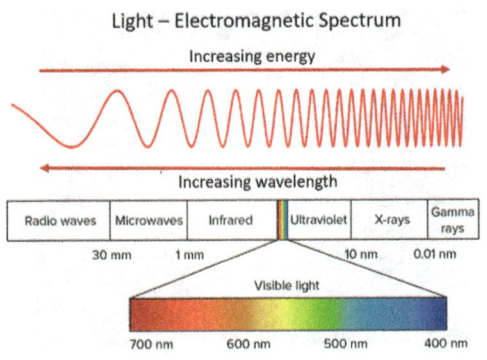

Visible Light

Visible rays of light that can be seen with our eyes are in the range of 3900-7700 Ao wavelengths. Where Ao is 10-9 mm in length.

In fact, there are an infinite number of colors depending on the wavelength of visible light. However, for convenience, visible light is divided into seven colors. These rainbow colors range from long to short wavelengths and are red, orange, yellow, green, blue, indigo, and violet. When the visible rays of these seven colors touch together, the color cannot be distinguished; therefore, it is called white light.

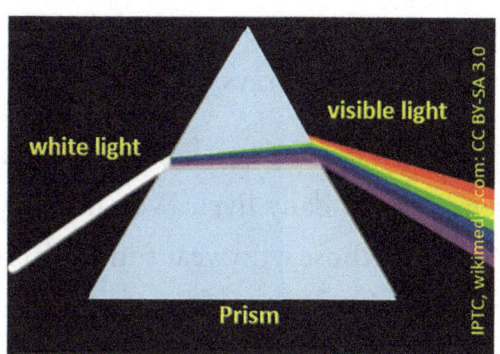

White light and Visible light

When we see an object, we are aware of the light that is reflected. For example, seeing a yellow-colored paper is not because the colored paper is yellow, but because the colored paper absorbs all other colors and reflects only the yellow wavelength. If we wear blue glasses and look at a piece of white paper, the reason we see blue is because the glasses pass through only the wavelength of blue light and absorb all other light. All of these phenomena can be recognized only in the visible light region.

Invisible Light

Outside of the visible light range, no one can detect light at all. Again, rays that cannot be seen are wavelengths outside the range of visible light. Examples of wavelengths longer than visible wavelengths include infrared, microwave, and radio waves. Short wavelengths include ultraviolet rays, X-rays, and gamma rays.

Although we cannot see these lights, we are aware of their existence in our daily lives. We can easily see the difference between those who wear sunblock cream versus those who don't after spending time on the beach. From this difference we know it's due to ultraviolet light (or UV light). Another example is that we cannot see the bones of our body with our naked eyes, but we can through X-rays.

Infrared rays are radiant energy that make us feel warm as if from a furnace. Heat is invisible, but it is energy that can be felt. The reason why an object has a red color when emitting heat is that the wavelength of infrared and the wavelength of red overlap. Through longer than infrared wavelengths, we receive radio broadcasts.

Fundamental Energy – Light

Since light is the basis of energy, it is very important that the Bible refers to how God first created light before mentioning the sun and the stars. Without light, the sun and the stars could not shine and could not imagine their own existence. The Bible implies that the light of the first day of creation is far more fundamental than the sun, moon, and stars, all of which appear on the fourth day of creation.

"How does light exist without the sun?"

This is the most frequently asked question from the audience. Many get the idea that the sun is sending light, and therefore the sun should have been made before light in the right order. However, in Genesis, we discover that God created light on the first day, and eventually on the fourth day, the great light (sun) and small light (moon) and the stars were created.

It is easy to see that light was created before the sun in the book of Genesis. Strictly speaking, the light and the sun are not the same things. As mentioned already, light is the basis of all electromagnetic energy that activates the motion of all molecules.

According to scientists, the sun is mainly composed of hydrogen and helium. Therefore, if the light did not already exist, that would indicate a serious problem with the role of the sun itself. Because we cannot imagine any chemical reactions of hydrogen molecules without light. In the next chapter, we will cover this further in detail. In that respect, it is remarkable that the Bible treats the light and the stars separately from the beginning of creation.

The light created on the first day is "ore" in the original Hebrew language. Meanwhile, the lights created on the fourth day are "maw ore." If we look at this in its original language, "ore" is the light itself. But "maw ore" is an object or device (light-giver) that produces light. In other words, on the first day, God made light, which is the source of energy, and on the fourth day, he made a device that emits that energy.

Electricity First or Light bulb First?

A simple example may help here. What would come first? Electricity or incandescent light bulbs?

Of course, electricity comes first! A light bulb is a device that converts its electrical energy into light energy. Therefore, it is very reasonable that light (ore) was created before the sun, as electrical energy precedes the light bulb.

If this order were written in reverse, the Bible would be more attacked by scientists. Who will follow the wisdom and omnipotence of a God who made the tool to harness light before the light itself?

Neither naturalistic evolution nor any scientific theories explain the existence and origin of light itself, but only the Bible gives the answer.

"Do you listen in on God's council? Do you limit wisdom to yourself?" (Jb15:8)

"Where is the wise man? Where is the scholar? Where is the philosopher of this age? Has not God made foolish the wisdom of the world?" (1Co1:20)

Chapter 8:
Light before the Sun

In the previous chapter, light and the sun were compared to electricity and a light bulb. As we discussed, which came first, electricity first or a light bulb? Of course, electricity came first! A light bulb is just a device that converts electrical energy into light energy; therefore, it is reasonable that light, the source of all energy, was created before the sun.

The Sun is not the Source of Light!

Physicists know that the sun is not the source of light. They understand that light energy is composed of photons, which are massless particles. That is why light is not regarded as matter.

No Light and No Matter!

In today's advanced particle physics, it is impossible to imagine matter without light. Scientists continue to learn through experiments that matter cannot be independent from light. [18] High-energy particle physics will be covered in more detail in Chapter 10.

THE UNIVERSE OF DYNAMICS

It was already introduced that the Bible refers to the creation of time-space-matter and light on the first day of creation. We don't have to mention the concept of continuum in physics, and we can see the depth of the words of the Bible.

In the Bible, God created light before creating the sun. With that light, he defined "day" and "night" (Gn1:5a). After that, evening and morning came, that is, although the sun was not yet created, he turned the earth so that it became the second day and third day. And then, on the fourth day, he made the sun, moon, and stars as light-bodies, and filled the space created on the first day (Gn1:14-18).

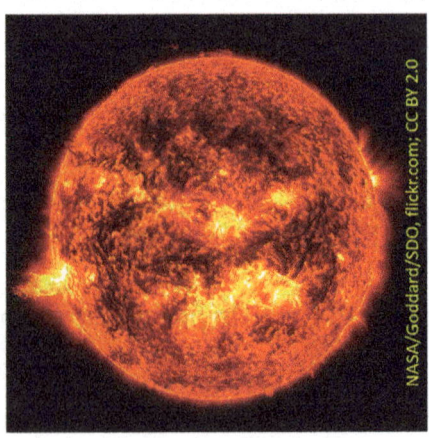

An Image of the Sun

Nuclear Power Plant – The Sun

Scientists who study the sun view the sun like a nuclear power plant. A nuclear power plant that supplies us with electricity controls the fission reaction of matter to produce electricity from the thermal energy generated when the nucleus breaks down.

Scientists believe that the heavier deuterium and tritium at the core of the sun undergo an endless fusion reaction, and as a result, turn into helium and emit light energy in the form of electromagnetic waves and neutrons. [20]

Here the hydrogen fusion includes the emission of neutrinos, positrons, and gamma rays as well. It should be noted that both fission and fusion are nuclear reactions that produce energy, but the applications are very different.

Fission or Fusion?

Fission reaction is the decaying (or splitting) of a heavy, unstable nucleus into two lighter nuclei. On the other hand, fusion reaction is the process where two light atoms (nuclei) combine together to form a heavier atom and release vast amounts of energy as shown in figure below.

So, fission is used in nuclear power reactors since it can be controlled, while fusion is not utilized to produce power since the reaction is not easily controlled and is expensive to create the needed conditions for fusion reaction. A lot of research on fusion is ongoing, but it remains in its experimental stages yet.

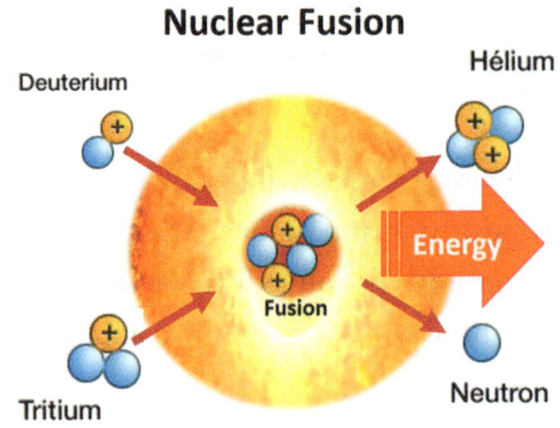

What's going on the Surface of the Sun?

For a long time, scientists have observed the phenomenon of huge cells of heat energy erupting from the sun's interior onto the surface. The size of a heat cell is about 1000 km in diameter. It pops up to the surface of the sun and moves rapidly at a speed of 7 km/sec. It drifts around the surface for about 20 minutes before it disappears. This hot heat energy forms the surface photosphere of the sun, the top of the convective layer, at 4,000 – 5,000 degrees Celsius.

Of course, even with advanced technology, no one can directly observe what is going on inside the sun. In general, it is understood that the temperature at the core of the sun is above 14 million degrees Celsius, and nuclear fusion reacts endlessly at the center. As shown in the diagram below, it is assumed that there

is a radiation layer at the core and a convective layer formed on top of it.

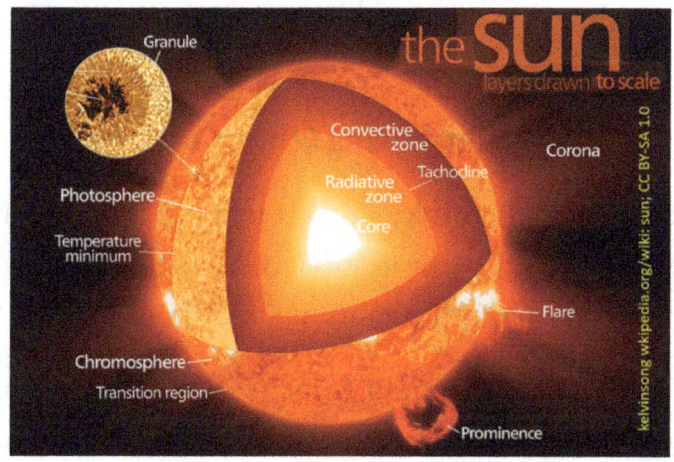

Cross-section Diagram of the Sun

No Chemical Reaction without Light?

In order for the fusion reaction to take place in the core of the sun, substances must undergo a chemical reaction with each other. However, it is impossible to conceive of a chemical reaction without light energy in the physical world. So, there must be light first.

If deuterium and tritium are sufficiently prepared in a nuclear power plant, the substances alone are not enough for the reaction to begin. It must be tuned to a specific pressure and temperature at which the fusion reaction can take place.

Substances cannot initiate reactions themselves without a specific input. Naturalists, of course, will bring a different story that this happened by chance, but this is a story far from experimental science.

The Sun is a Light-Giving Object

According to the Bible, God created light on the first day of creation. On the fourth day, he made the sun as a light-giving object that emits light. [20]

Therefore, the account of Genesis in the Bible is very straightforward, scientific, and convincing. In the next chapter, we will share a couple of basic laws governing the matter world, which can be called "the order of creation."

"How many are your works, O LORD! In wisdom you made them all; the Earth is full of your creatures." (Ps 104:24)

"Then I saw all that God has done. No one can comprehend what goes on under the sun. Despite all his efforts to search it out, man cannot discover its meaning. Even if a wise man claims he knows, he cannot really comprehend it." (Ec 8:17)

Chapter 9:
Laws of Creation

Building Blocks

Chemistry is a field that deals with matter and energy and their interactions. In chemistry, elements are building blocks that determine all matter in the universe.

An element usually refers to a substance that can no longer be divided into smaller substances. So far, 118 elements have been discovered, of which 94 are naturally occurring.

Scientists are describing all matters in the universe with these elements and their combinations. Interestingly, all scientists are saying that these elements are the building blocks of the universe. They are the building blocks of the big house!

Matter is Energy!

Matter is anything that has mass occupying certain space. Energy is the potential to make a change. In other words, energy is the ability to do work. It has the potential to give order to all things.

Light, heat, sound, etc., are all forms of energy. It tells us that energy was acted upon when a change occurred in the matter world. There is change whenever one form of energy is converted into another.

Even a Tiny Dust is a Massive Energy!

Einstein found that the energy of matter is proportional to the mass of matter and the square of the speed of light. Briefly speaking, matter and energy can be mutually exchanged. Even a tiny speck of dust has a huge amount of energy. Since energy is proportional to the square of the speed of light, it is now possible for engineers to calculate the power of an atomic bomb as a numerical value.

Put simply, matter is a mass of energy; therefore, this matter world is nothing but a humongous ball of energy. However, the matter world is governed by strict laws. Now let's consider very fundamental natural laws in science.

The First Law

The first law we should note is the law of conservation of energy. Again, energy can appear in different forms, such as heat, light, sound, chemical, kinetic, or electrical energy.

For example, an oven in the kitchen converts electrical energy into heat energy, a light bulb converts electrical energy into light energy, and a car engine burns gasoline fuel to convert chemical energy into mechanical energy. Thus, it is clear that energy can be transformed from one form to another but can neither be created nor destroyed. This is the first law of thermodynamics.

Then, a natural question may arise here. How did the existing energy or matter come into existence? This is the most fundamental question in understanding the material world. However, no reference books can answer this question. Only the book of Genesis in the Bible teaches creation from absolutely nothing from the beginning.

The Second Law

The second law governing the physical world is the law of increasing disorder. This law is known as the second law of thermodynamics. The first law relates to the total amount of energy, and this second law relates to the quality of energy.

According to both laws, the total amount of energy is constant without increasing or decreasing in the universe; however, useful energy decreases, and unusable energy increases over time. In other words, the 2nd law states that the degree of disorder increases over time. Thus, the order of all matter in the universe is gradually declining.

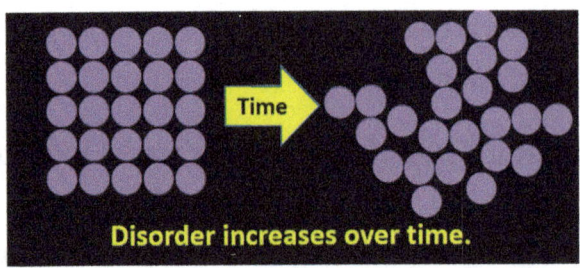
Disorder increases over time.

More precisely speaking, the 2nd law of thermodynamics postulates that the entropy (or a measure of disorder) of a closed system will always increase with time. Of course, the entire universe is an isolated system because it has no surrounding. Hence the universe is the only known closed system. Note that living organisms are not a closed system, and therefore the energy input and output of biological organisms are not directly applicable to this law.

Anyway, because of this 2nd law, the products that have been exhibited at the factory cannot be renewed, and the food left over will rot away soon.

Everything is declining! Everything that exists in nature is ruled by this law. No exception! If exceptions exist, they may be evidence of evolution, but there is no single exception.

In order for evolution to occur, it must proceed in the direction of getting better or more organized over time, but this law completely rejects the theory of evolution.

Meaning of the Laws

The above two laws are laws of nature. These laws are the most universal and best-proved generalizations of science. They are applicable to every process and system of any kind. Like Newton's law of motion, these thermodynamic laws are also the laws of causality that can be explained by cause and effect.

Let's briefly consider the meaning of these laws. The first law is that in the beginning the universe was created out of nothing. The second law is the teaching that this universe had a beginning and someday will end.

The Bible clearly teaches that there was a beginning, and the beginning was good. In particular, Genesis chapter one repeats seven times that 'the first was good.' No book in the world teaches that the first was good. Only the Bible!

THE UNIVERSE OF DYNAMICS

The two laws introduced above are the most basic in science. Through these laws we learn that no matter how simple and tiny a speck of dust is, nothing in the universe has been created by accident. Even the most outstanding scientist cannot produce electronic dust in a perfect vacuum laboratory.

We have looked at just two laws, but there are many more mysterious laws in nature. All the laws are the product of the Creator's wisdom. Please note that all scientific laws demonstrated by experiments support the Bible. In other words, there is no conflict between experimental science and the Bible.

We know that many more laws governing the world are coordinated with each other, so we cannot deny the eternal power and divine nature of the creation, as in Romans 1:20. Representatively, the Old Testament Nehemiah (9:6) says that "God created and preserves all things." In this way, the Bible testifies that all things in the universe are being harmonized and preserved by the strict laws of the Creator's order.

"Oh, the depth of the riches of the wisdom and knowledge of God! How unsearchable his judgments, and his paths beyond tracing out!" (Rm 11:33)

Chapter 10:
What is the End of Matter?

If we come close to look at a sandy beach or a sedimentary sandstone layer in a canyon, we can see that it is made of tiny hard pieces of sand. We can touch some of sand grains on the beach or the sandstone.

Suppose we put a small grain of sand on a very hard base and hit it with a hammer. Wouldn't we smash it into smaller pieces? Of course, we could smash one of those smaller pieces into still further smaller pieces. What if we could keep on doing this forever? If we keep on doing that, what will be the end?

THE UNIVERSE OF DYNAMICS

Earlier Greek scholars, Leucippus (450 BC) and his pupil, Democritus (460-370 BC) had such questions. They began to think not only rudimentarily but also systematically about the ultimate end of matter. They thought that if they broke a piece of an object into smaller pieces, broke the smaller pieces into even smaller pieces, and then repeated this process, then one day they would reach the end of matter so that it could no longer be broken down. Later, Democritus called the ultimate kernel (or core), which could no longer be divided, 'Atomos' in Greek. Today, this is called 'atom.'

Democritus thought the whole world was made up of different kinds of atoms and that in between the atoms there was nothing at all. An atom is too small to be seen, but when many of them were joined in different combinations, they made up all the different things we see around us. Also, he thought atoms could not be made or destroyed, although they could change their arrangements. In such a way, one substance would be changed into another. [22]

This concept of matter has evolved over history, and later scientists say that the universe is made up of tiny particles atoms. We all believe that we are made up of atoms too. An atom is the smallest particle of a substance that still has all the qualities of that substance.

Until the nineteen century, people thought that the atoms could no longer be broken down. With the continued progress of science and technology, scientists realized that atoms are made of even smaller particles. It was found that an atom has a nucleus in the middle, and electrons around it. In the center of each atom, protons and neutrons are strongly combined in the nucleus. It is the periodic table of elements in chemistry textbooks that classify and organize all elements according to this atomic model.

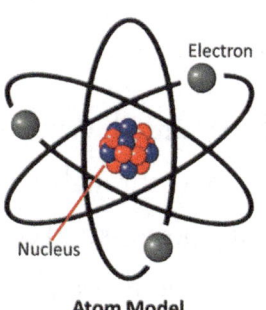

Atom Model

These days, however, the up-to-date atomic model is the electron-cloud model as shown in the figure below. The nucleus is the atom's dense center. The cloud around the nucleus contains the tiny electrons. And these electrons appear from place to place around the nucleus in the electron cloud. [23]

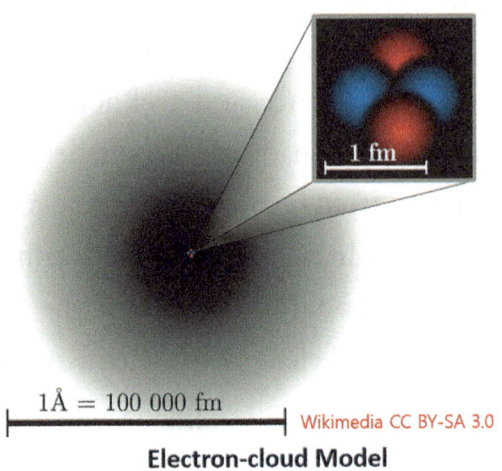

Electron-cloud Model

Atomic Particles

In the middle of the twentieth century, particle accelerators were developed, and through technology, scientists succeeded in separating protons and electrons. This reveals the problem of the atomic theory, which suggested that atoms could not be broken up any further. Subsequently, through advanced particle accelerators, the atomic nuclei were separated into protons and neutrons again.

Experiments in which proton beams accelerated with high energy collide into each other, making it possible to break the protons into smaller pieces. As a result, the particles split from the collision experiment of the proton beams are called "quarks."

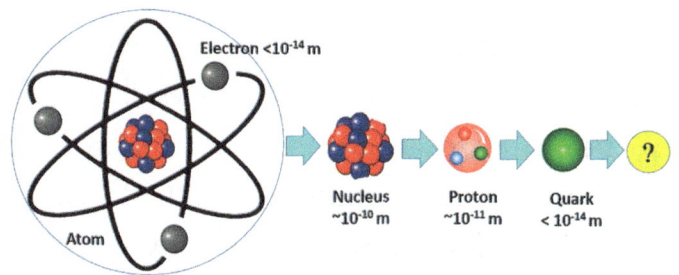

A model of Atom-Nucleus-Proton-Quark

In the field of particle physics, the standard model suggested by scientists as a result of research is shown in the picture below.

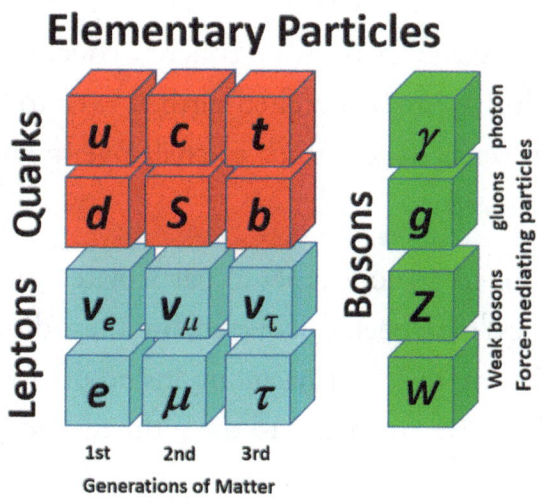

There are 6 quark particles, 2 boson particles that mediate their forces, 6 lepton particles, and 2 boson particles that also mediate their forces, all of which mean there are 16 elementary particles.

It was expected that there would be another particle that gives these 16 elementary particles their mass. This particle is the socalled God particle, or the Higgs boson particle. It wasn't until July 4, 2012, that the European Council for Nuclear Research (CERN) laboratory announced they had found the Higgs boson particle.

Labyrinth of Matter-End

Since its discovery, research on the Higgs particles is of great interest to the scientific community and is in full swing these days. There are no concrete conclusions yet. The results have to be watched further. Even today, cutting-edge scientists continue to challenge the ultimate source of matter.

As mentioned, various types of particulates (elementary particles) have been discovered, but these discoveries reveal the mystery and harmony inside of the matter rather than give an answer to the ultimate end (origin) of the matter. [24-25]

Here some scientists may be talking about quantum theories or mechanics for explaining the behavior of subatomic systems. Note that a quantum is a discrete quantity of energy proportional in magnitude to the frequency of radiation.

Quantum mechanics studies the motion and energy of quanta, which includes electrons, protons, neutrons, photons, etc. However, until now, research on quantum mechanics has not reached a definitive conclusion. Meanwhile, the results of various advanced research have consistently shown mysterious behavior and uncertainties of quanta.

As long as science and technology develop, such research will continue to progress endlessly until they give up the study of breaking particles because of the limits of technology. Likewise, the end of the universe, the ultimate end of matter, seems to be a labyrinth as well.

Unexpected Test Results

Those who take a naturalistic position believe that the universe started from a tiny "kernel," and galaxies arose, from which solar systems, planets like the earth, and also life accidentally appeared, as we witness. Therefore, they believe that all origins are from simple matters.

Again, they think it evolves from the simplest matter to the most complex. This is the position of naturalistic evolution. That's a plausible claim! But what we learned through various experiments is the opposite. It is more unstable when simple particles exist independently.

Simpler is less stable!

For example, the neutrons that make up an atomic nucleus are the most unstable particles. When they are separated from the nucleus, they decay in only 10.2 minutes. Therefore, a bonded state with protons is more stable than when the neutrons are independent.

Although free protons are remarkably stable, generally speaking, a proton is more stable than the quarks, an atom is more stable than the nucleus, and the molecule is more stable than the atom. This is contrary to the position of evolutionary concept.

Scientific Findings

The results of the research so far can be briefly summarized as: 1) all matter is made up of particles called atoms. 2) An atom has protons and neutrons bound to its nucleus and has electrons around it. 3) Atoms cannot be spontaneously created or destroyed by themselves. 4) Particles smaller than an atom are unstable when they are independent.

Thanks to today's state-of-the-art research, it has been found that the smallest and simplest particles cannot arise by themselves, and a tiny particle cannot be explained from the point of view of chance and spontaneity. Through the particle world, it became possible to see inevitable order and harmony. Therefore, we became more aware of the mystery and power of God's creation.

"Then I saw all that God has done. No one can comprehend what goes on under the sun. Despite all his efforts to search it out, man cannot discover its meaning. Even if a wise man claims he knows, he cannot really comprehend it." (Ec8:17 NIV)

"Oh, the depth of the riches of the wisdom and knowledge of God! How unsearchable his judgments, and his paths beyond tracing out!" (Rm11:33 NIV)

Chapter 11:
Spinning Earth from the Beginning

"And there was evening, and there was morning—the first day." (Gn1:5b)

"How could a day be without the sun?"

This question naturally comes up when we read Genesis chapter one. If the Earth was created on the first day and the sun on the fourth day, then is it not contradictory to the Bible from the first to the third day?

Some people claim that one day in the week of creation is not equivalent to a day in current times, but an unknown period. To such people, we should ask the following question.

"How does a day become a day?"
Then, they may probably answer like this:
"Because the sun rises in the east and sets in the west."

Truly, it is a celestial answer! In this era, no one believes that the sun moves around the Earth. Nevertheless, we still think that when the sun orbits our Earth it makes a day. However, we all know that one day is not caused by the sun, but entirely due to our Earth's rotation. In that respect, it is very surprising that the Bible used the time unit of one day on Earth alone before the sun was created.

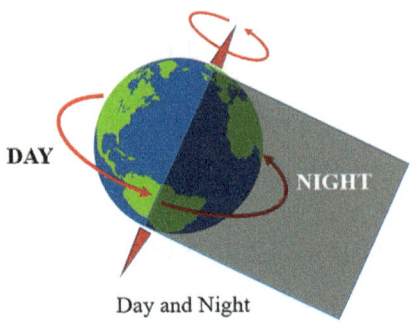

Day and Night

Time and Cycle of Time

In the first verse of Genesis, God created "time" called "beginning." He not only created time, but also created "a period of time."

The first unit of time is a day. This "day" appears without the sun and only after the Earth was created. In other words, God made the Earth rotate from the beginning and set a cycle of one day with the Earth alone.

What if God didn't use "day" from the first to the third day before the sun, but used the word "day" from the fourth day when the sun was created? The Bible would have been criticized even more.

A fact is true, and it cannot be changed. A "fact" does not change even if scientific theories change. If astronomers in the past said that the sun orbits around the Earth, the Earth still orbits the sun. Therefore, the Bible in which facts are written does not change, nor does it need to be changed.

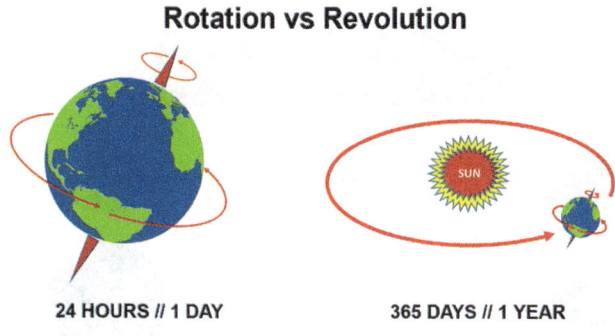

God created another period of time, a "year." This "year" first appears when the sun was created on the fourth day (Gn 1:14).

As we have seen in the solar system, everything in outer space revolves around heavier celestial bodies. If two celestial objects have the same mass (or weight) in space, then they will orbit each other, and the middle point will be the center of their orbit as shown in the diagram below. If they have different masses, then their orbit center will be shifted somewhere close to the bigger mass depending on the ratio of their masses.

With the same principle, the sun would also orbit as much as the mass ratio with respect to each planet, but the mass of the sun is so dominant that it seems like it is not orbiting.

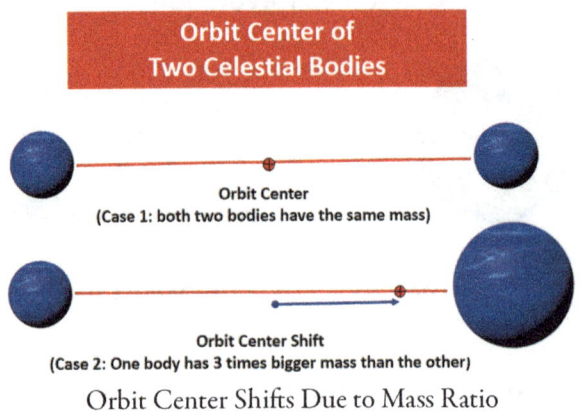

Orbit Center Shifts Due to Mass Ratio

If we apply this principle, now the order of "day" and "year" written in the Bible can be easily understood. Since the earth rotated three times—the first day, second day, third day, and during the fourth rotation, a much more massive sun was created, so the earth immediately began to orbit the sun. The Bible mentioned exactly the "year." It is the perfect order! After creating two light-giving bodies—the sun and moon, God said, "Let them serve as signs to mark seasons and days and years" (Gn1:14b).

What Makes a Day?

Before Copernicus, no one would have thought that the earth's rotation is what causes the "day." In other words, it was not possible to conceive of the earth rotating and moving around the sun. The reason is very simple. Because no one has ever seen the planets, including the earth, from outside of the solar system. Also, apparently, the sun rises from the east and moves to the west.

The Bible has been saying from the time of creation that the "period of time" can only be explained through the rotation and orbit of the earth. These Genesis records cannot be conceived except for the Creator who can consider time, the unit of time, the earth, the sun, and gravity, all things at once. Isn't it as if God is speaking directly from outside of space beyond the solar system and galaxies, looking at the earth, the sun, the planets, and the stars!

"Have you ever given orders to the morning, or shown the dawn its place?' (Jb 38:12 NIV)

"Hast thou commanded the morning since thy days: and caused the dayspring to know his place?" (Jb 38:12 KJV)

This question God posed to Job is very meaningful! Because when the first morning was there, no one was there yet.

We can say there are two kinds of people in the world. One kind of person ponders alone how the morning began. The other kind finds and asks who created the morning and looks for witnesses who were there. This question to Job is saying that he is the Creator of the morning. He said of the first morning: "Evening and morning were the first day."

A Blurry Start or a Clear Start?

Our beginning is not dark or vague, but rather it is very clear because we know the Almighty God who transcends the morning, and He was there in the beginning. With naturalism or evolution, people think that what is seen is everything, all beginnings, including the morning, are only blurry and vague.

He not only created, but has maintained everything since creation, including all the planets of the solar system and all the stars in the universe. The Bible tells us that he was not the one who thought he had finished his mission as creation, but that he is still working after that. The Bible testifies that he is the Creator and Jesus. In the next chapter, we continue to talk about the Creator and his will.

Chapter 12:
Just and Love

Through each chapter, we have come to see how dynamic and mysterious this universe is. We have seen that the entire universe is truly a dynamic world, not chaos, but cosmos! From a dynamics perspective, the solar system is the result of inevitability, and the motion of the Galaxy is an intentional consequence as well.

At this very moment, thousands of satellites are orbiting the earth that were made by humans and set to revolve at the specific orbit of the earth. In a solar system of an unimaginable scale, it has been confirmed that all of the planets and their satellites orbit the sun on the same principle as a man-made spacecraft. Each performs a unique motion in space. No one will deny that this universe is dynamic.

Who set up this humongous and precision dynamic system?

From a dynamic perspective, we have come to a clear conclusion. The universe is created by wisdom.

THE UNIVERSE OF DYNAMICS

The Scripture that testifies of creation is a record that cannot be reached by human wisdom. The Bible provides clear answers to problems that science cannot give. It has been found that the scientific findings and verified data do not contradict with the Bible.

The Bible Tells Facts

The Bible and experimental science do not conflict with each other. No worldly book provides a definitive answer to questions related to the origins of the universe. Only the Bible gives us direct answers.

The Bible is God's word, and it testifies that all are pure and flawless (Pr30:5, Sm22:31, Ps10:30). Apostle Paul says, "All Scripture is inspired by God..." (2Tm3:16a). All these verses tell us that the Scriptures did not come from human wisdom. It is the Scripture that God moved the chosen people and wrote his words by their hands. (Note that the Bible was written on the same subject by 40 different people over 1600 years.)

The Word and God

The Bible begins with the proclamation of the creation of the world. God is the Creator, and the creation was done by Him. God used to mention "God said" whenever he created something in Genesis chapter one. This implies that God created all things with his Word.

In the New Testament, Apostle John states: "In the beginning there was the Word, and the Word was with God, and the Word was God." This implies that God and the Word are one, and we cannot think of the two separately.

Apostle John professed: "The Word became flesh and made his dwelling among us. We have seen his glory, the glory of the One and Only, who came from the Father, full of grace and truth" (Jn 1:14). He said that the Word came to this earth in the flesh to dwell with us. This is the only begotten Son, Jesus Christ.

The Bible testifies about Jesus:

"The Son is the radiance of God's glory and the exact representation of his being, sustaining all things by his powerful word. After he had provided purification for sins," (Heb 1:3a)

According to this passage, His Son, Jesus, is the reality of God and the Creator. He is the One who can hold all things with his Word, and only he can cleanse our sins.

If he is not the Creator, how can he cleanse our sins and be our Savior? The Bible consistently declares Jesus as the Savior who himself paid the penalty for our sins and affirmed the love of God.

The Creator Jesus on Earth

"He was with God in the beginning. Through him all things were made; without him nothing was made that has been made. All things were made by him; and without him was not anything made that was made." (Jn 1:2-3)

The Bible continuously states that "He was in the world, and though the world was made through him, the world did not recognize him. He came to that which was his own, but his own people did not receive him." (Jn 1:10-11)

Jesus came to this Earth as a perfect man and was 100% human. The Bible tells us that he, the Creator and Master of the world, came to this world in the flesh. Through his actions that he showed during his lifetime, we can know that he is God and his will is God's will.

The lectures of Jesus are found in Matthew chapters 5 through 7. His teaching is known as the Sermon on the Mount. After his teaching, Jesus then came down from the mountain, and showed people how to apply his teachings directly to their lives. His deeds were described in detail in Matthew chapters 8 and 9. The ten miracles written in these two chapters are clear examples that show that he is the Almighty Creator. Obviously, he transcends the physical world and living world as well.

Just and Love – The will of God

If God created the world, what is his intention? Why did he, who is omnipotent, come to this world?

"I am the Lord your God; consecrate yourselves and be holy, because I am holy...be holy, because I am holy" (LV11:44a, 45b).

"It is God's will that you should be sanctified." (1Th4:3a)

This is the will of God. To accomplish his will God sent Jesus Christ to this Earth.

In fact, the entire Bible testifies of Jesus. The first four books of the New Testament—Matthew, Mark, Luke, and John—are the history of Jesus' life. The following verse written by Apostle John represents God's will well.

"For God so loved the world that he gave his one and only Son, that whoever believes in him shall not perish but have eternal life." (Jn3:16)

Jesus himself said: "For I have come down from heaven not to do my will but to do the will of him who sent me." (Jn6:38) And he continued; "I have come that they may have life, and that they might have it more abundantly." (Jn10:10b) "Just as the Son of Man did not come to be served, but to serve, and to give his life as a ransom for many."(Mt20:28, Mk10:45)

Among the disciples, the one who experienced the love of Jesus the closest is Apostle John. He confessed in the book of 1 John: "Dear friends, let us love one another, for love comes from God. Everyone who loves has been born of God and knows God. Whoever does not love does not know God, because God is love." (1Jn4:7-8)

And he continuously testified his love by saying: "This is love: not that we loved God, but that he loved us and sent his Son as an atoning sacrifice for our sins." (1Jn4:10)

Jesus prayed at the Last Supper the day before his death on the cross: "I in them and you in me. May they be brought to complete unity to let the world know that you sent me and have loved them even as you have loved me." (Jn17:23) In such a way, he wanted his followers to know that our heavenly Father God loves them as much as the Son himself.

This is because we are truly formed in his image. That is why the Old Testament Zephaniah sings: "The LORD your God is with you, he is mighty to save. He will take great delight in you, he will quiet you with his love, he will rejoice over you with singing" (Zph3:17).

Therefore, God the Father showed his infinite love: "He who did not spare his own Son, but gave him up for us all—how will he not also, along with him, graciously give us all things?" (Rm8:32)

God is holy and just. He loves us infinitely, but he cannot be with our sins. God sent the innocent Jesus Christ to this world as a peace offering to save this fallen world. He took the sins of the world, died on the cross, and was resurrected. He paid the penalty of our sin on the cross. God's justice was accomplished. Those who believe in him will be free from the chains of sin and be saved. It is the good news, gospel!

And "but God demonstrates his own love for us in this: While we were still sinners, Christ died for us" (Rm5:8).

The Bible tells us that God sent Jesus to the earth because of his love. His son hung on the cross to fulfill that love, confirming God's love and completing his justice. No one can deny God's justice.

Authority of Faith

Faith is defined as "complete trust" or "confidence in someone or something." The Hebrew writer defines faith like this: "Now faith is being sure of what we hope for and certain of what we do not see" (Heb 11:1). Apostle Paul says, "So then faith cometh by hearing, and hearing by the word of God " (Rm 10:17 KJV). Accordingly, faith is to believe the word of God, the Bible.

Since the Bible was written, the Bible has been the same. This is because the Bible is true, and the facts cannot be changed. The highest authority of Christian faith goes to the Bible. No matter how knowledgeable, experienced, and accomplished. Regardless of what our beliefs and practices are, it does not matter. The ultimate authority of our faith is the Bible and if there is anything higher than the words of the Bible, it is labeled as an idol.

Faith and Deeds

The Bible is not intended to be understood by only a specific group or people. It is written for everyone. Proverbs of the Old Testament says: "My mouth speaks what is true, for my lips detest wickedness. All the words of my mouth are just; none of them is crooked or perverse. To the discerning all of them are right; they are faultless to those who have knowledge" (Pr8:7-9). This means that everyone who has common sense, from children to the elderly, can understand.

Apostle Paul tells us about faith and salvation through the following verses:

"For it is with your heart that you believe and are justified, and it is with your mouth that you confess and are saved." (Rm10:10)

And "For it is by grace you have been saved, through faith—and this not from yourselves, it is the gift of God." (Eph 2:8)

According to these verses, faith is not obtained through our hard work or effort. It is not obtained by our actions or qualifications, but rather it is God's totally free gift. Faith is grace, and faith is what is believed. The Bible consistently teaches that grace is a free gift from God. No matter what your faith and practice are, it does not matter. It is free! At the same time, the Apostle Paul says in verse 10 that we come to salvation by admitting with our mouths. This also teaches the importance of volitional confession.

THE UNIVERSE OF DYNAMICS

"Yet to all who received him, to those who believed in his name, he gave the right to become children of God." (Jn1:12)

"If you hold to my teaching, you are really my disciples. Then you will know the truth, and the truth will set you free."(Jn8:31b-32)

Finally, I deeply thank all of you who have read this book. I wish you to confirm through the Bible that you are a more precious image of God than anything else, including the universe. Enjoy true freedom with the truth of God. His grace and peace be with you! Thank you and God bless!

References

[1] Newton, I. General Scholium. Translated by Motte, A. 1825. *Newton's Principia: The Mathematical Principles of Natural Philosophy*. New York: Daniel Adee, 501. The Greek word pantokrator is most often translated as "Almighty" in the King James Version.

[2] Moody Science DVD, *Wonders of God's Creation: Planet Earth*, Moody Publishers, 1993.

[3] Cameron, A.G.W., *Abundances of the elements in the solar system*, Space Sci Rev **15**:121-146, 1973.

[4] Anders, E. and Ebihara, M., *Solar-system abundances of the elements*, Geochim. Cosmochim. Acta **46**:2363-2380, 1982.

[5] Murray, M.J., *Reason for the Hope within* (Grand Rapids, MI: Eerdmans), 1999.

[6] Strobel, L., *The Case for a Creator: A Journalist Investigates Scientific Evidence That Points Toward God*, Zondervan, Grand Rapids, MI., 2004

[7] DeYoung, D., *Astronomy and the Bible*, Baker Book House, 2010

[8] Scott et al.; in: Cox, A.N., *Allen's Astrophysical Quantities*, 4th Edition, Springer-Verlag, New York, 2000.

[9] Milky Way: *Wikipedia Encyclopedia*, Edited on 8 March 2021.

[10] Morris, H., *Henry Morris Study Bible Apologetics Commentary and Explanatory Notes*, Master Books, 2014.

[11] Humphreys, D.R., *Our Galaxy is the center of the universe, "Quantized" redshifts show*, TJ **16**(2): 95-104, August 2002.

[12] Sarfati, J., *The Sun: Our Special Star*, Creation **22**(1): 27-30, 1999.

[13] Chown, M., *What a star!* New Scientist **162**(2192): 17, 1999.

[14] Morris, M., *What's happening at the center of our galaxy?* Physics World (Oct. 1994), pp.37-43, 1994.

[15] St. Augustine (Bishop of Hippo), *Confessions*, Hackett Publishing, 2006

[16] Humphreys, D. R., *Starlight and Time*, Master Books, 1998.

[17] Morris, H., *Henry Morris Study Bible – Apologetics Commentary and Explanatory Notes*, Master Books, 2014.

[18] Schine, N. and et al, *Synthetic Landau Levels for Photons*, Nature 534, 671-675, 2016.

[19] Rainwater, C., Light and Color, Golden Press. New York, 1971.

[20] The Sun as a nuclear reactor: http://library.thinkquest.org/ c001124/gather/ssun.html, 2006.

[21] Henry, J., *The Astronomy Book*, pp.40-43, Master Books, Green Forest, 1999

[22] Asimov, I., *How did We Find out about Atoms?* Walker and Company, New York, 1976.

[23] Cohen et al, *Discover Science*, Scott, Foresman and Company, 1991.

[24] Perkins, D. H., *Introduction to High Energy Physics*, Cambridge University Press, 4th edition, 2012.

[25] Veltman, M., *Facts and Mysteries in Elementary Particle Physics*, World Scientific Publishing, 2018

Additional References

[26] DeYoung, D. B., *Astronomy and Creation An Introduction*, Creation Research Society Books, 1995.

[27] DeYoung D. B., *Science and the Bible*, Vol.2, Baker Books, 2002.

[28] Whitcomb, J. C., *The Bible and Astronomy*, BMH Books, Winona Lake, IN, 1984.

[29] Faulkner, D., *Universe by Design – An Explanation of Cosmology and Creation*, Master Books, 2006.

[30] Gitt, W., *Time and Eternity*, Loizeaux Neptune, NJ, CLV, 2001.

[31] Gitt, W., *Stars and their Purpose*, Signposts in Space, CLV, 2000.

[32] Gitt, W., *In the Beginning was Information*, CLV, 3rd edition, 2001.

[33] Lisle, J., *Astronomy – A Pocket Guide, What is the Biblical Perspective?* Answers in Genesis, 2018.

[34] Lisle, J., *Taking Back Astronomy – The Heavens Declare Creation*, Master Books, 2006.

[35] Morris, J., *Is the Big Bang Biblical? And 99 Other Questions*, Master Books, 2003.

[36] Mulfinger, G., *Design and Origins in Astronomy*, Creation Research Society Books, 1989.

[37] Ashton, J., F., *In Six Days – Why 50 Scientists Choose to Believe in Creation*, Master Books, 2002.

[38] Brown, W., *In the Beginning: Compelling Evidence for Creation and the Flood*, Center for Scientific Creation, 1995.

[39] Slusher, H., S., *The Origin of the Universe: An Examination of the Big Bang and Steady State Cosmologies*, Institute for Creation Research, 1980.

[40] Ham et al, *The New Answers Books*, Master Books, 2012.

[41] Ham et al, *War of the Worldviews – Powerful Answers for An Evolutionized Culture*, Answers in Genesis, 2006

[42] MacArthur, J., *The Battle for the Beginning: The Bible on Creation and the Fall of Adam*, Thomas Nelson, 2001.

Recommended Websites

Creation Research Society: www.CreationResearch.org

Answers in Genesis Ministry: www.AnswersinGenesis.org

Institute for Creation Research: www.ICR.org

Association for Creation Truth: www.HisArk.com